Patenting for Enforcement

Designing in Patents

Practical Patent Strategies Used by Successful Companies

Eric Stasik,
author of Patent or Perish

Developing Other IP

Avoiding Rent

Advantages of Trade Secret Protection

Strategic Licensing

Excerpted from Patent or Perish: A Guide for Gaining and Maintaining Competitive Advantage in the Knowledge Economy

(includes expanded and updated information)

PATENT 08 • Stockholm, S.E. •Patent Engineering, Development, and Licensing Services

ALTHOS Publishing

About the Author

 Eric Stasik is director of Patent08.com, an expert consulting firm providing patent engineering, product development, and licensing services to small and medium sized businesses, individuals, and research organizations who want to creates strong new patent positions and maximize the leverage available from their existing portfolios.

Formerly director of patents and licensing for telecommunications giant LM Ericsson, he has over a decade of experience in dealing with complex patent issues and was instrumental in developing a patent portfolio which has generated in excess of €100 million in royalty income. Mr. Stasik has a Master's Degree in Electrical Engineering from the Georgia Institute of Technology in Atlanta, Georgia USA and has been a registered U.S. patent agent since 1994.

Table of Contents

Introduction

The basic challenge facing any company in a market economy is to use innovation to create positions of competitive advantage. In many industries, however, advanced technology renders these hard-won positions vulnerable to duplication, imitation, and plain, old-fashioned pirate copying which undermines competitive advantage almost as fast as it is created. In such an environment, patent protection, the limited monopoly which grants innovators time to commercially exploit or license their innovations, is often the best - and sometimes the only - way in which companies and individuals can profit from their innovations. Patents can make, or break, a business plan.

Despite the importance to business planning and success, patent strategies are often presented as abstract legal or economic concepts which provide little or no value to business and technology managers who are not patent experts. Examining how patents are used by leading companies in a specific business application can provide great insight to their practical use and application in your business plan. Unfortunately, most patent strategies are obscured by complexity and secrecy which makes examination difficult.

This document, which is sourced entirely from publicly available information, describes certain aspects of how Hewlett Packard and Lexmark have used patents to create the tremendously profitable ink-jet printer business. The example of ink jet printer cartridges provides a useful road-map to show how patents may be used in other technologies to create similar advantages.

Herein you will find a presentation of the basic elements of a patent strategy, how these basic elements are applied in the specific case of ink jet printer cartridges, and a discussion of how products and patents can be engineered to ensure that competitive advantage is gained - and maintained.

This is by no means a detailed, or comprehensive, study of ink jet printer technology. Instead certain aspects are highlighted to illustrate the practical use of patents.

Chapter 1 - The Business Model

Anyone who purchases a replacement ink-jet cartridge for a thermal ink-jet printer quickly understands that the business model of computer printer manufacturers like Hewlett Packard (HP) and Lexmark is to sell customers a high quality, yet rather inexpensive, computer printer which requires them to buy outrageously expensive replacement ink-jet cartridges from the printer manufacturer.

Offering consumers a low cost of entry in order to bind them to a high cost of "aftermarket" consumables (parts, service, supplies, accessories, or any combination thereof) is a successful business model used to sell products as diverse as automobiles, razor blades, jet engines, and computer printers. Ironically it was a socialist, King Camp Gillette who created what is one of capitalism's most successful business models.

"*Invent something people use and throw away,*" was the advice Gillette received in 1895 from his boss, the man who invented cork-lined bottle caps.[i] And he did: in 1901 Gillette patented the first safety razor to use disposable blades. While undoubtedly noteworthy in its own right, Gillette's safety razor was simply the physical embodiment of an even more remarkable idea. Not only did he "*invent something people use and throw away,*" Gillette devised a business strategy which locked his customers into buying all of their replacements from him.

Gillette's basic business strategy was to use a primary product, the safety razor, to sell consumables, disposable razor blades. His business innovation was to combine purposeful product design, aggressive marketing, attractive pricing, and the sagacious use of patents to lock-in his customers and build an installed base.

An Installed Base

The success of Gillette's model was based on building an "installed base" of steady customers. With the assumption that each safety razor owner would buy at least one new blade per month, 12 per year, the annual volume of consumables sold increased at a rate compounded by the sale of the primary product. In other words, as long as customers were locked-in to buying their replacement blades from him, even when sales of new safety razors were flat, the installed base would continue to grow.[ii] The (ideal) relationship between primary product sales and the growth of the installed based is illustrated below in Table 1.

	Year 1	Year 2	Year 3	Year 4	Year 5
Sales of Primary Product (Units)	1000	1000	1000	1000	1000
Installed Base of Users	1000	2000	3000	4000	5000
Annual Sales of Consumables (Units)	12000	24000	36000	48000	60000

Table 1 shows the (ideal) relationship between primary product sales, growth of the installed base of users, and growth in the sale of consumables.

Profit Margins which Grow Faster than Sales Volumes

As the size of the installed base grows and the volume of consumables sold increases, efficiencies in production, manufacturing, and distribution (i.e., "economies of scale") are gained which increase unit margins substantially. The result is a growth of profitability over time which exceeds even the growth of sales.

Low Volatility

Another appealing aspect of this model is that sales of consumables are generally less volatile than sales of primary products. Not everyone may replace their safety razor once a year, but nearly everyone who already owns one will buy a dozen replacement blades. The larger the installed base, the less sensitive profits will be to customer churn.

Persistence

The model also provides a measure of persistence in future sales. Depending on the physical and technical durability of the primary product, even after the sale of it is discontinued, sales of consumables may continue for years onward.

Product and Pricing Strategy

Product and price strategy form an integral part of the overall business strategy. Since every primary replacement is an opportunity for a customer to switch to a competitor's product, the primary product must be durable. In general, businesses with aftermarket sales compete aggressively with each other on price for new customers. Occasionally the primary product is sold at a deeply discounted price to acquire an installed base of customers who can subsequently be milked for replacement parts, maintenance, service, or repair. There are some risks associated with this. Customers happy to pay €5 for a razor handle may be reluctant to pay €2 for a replacement blade. Moreover pricing the primary products too low lowers the costs of switching and invites churn.

Locking-In the Customer

The largest risk to this business model occurs if users are able to buy their consumables from competitors or other third party suppliers. The whole strategy falls apart if the consumables are allowed to become commodities. Customers must be "locked-in" to a particular brand, or model, for which there can be little, or no, direct competition.

Typical strategies for locking in customers include creating service contracts, providing service expertise, strategic use of technology, and enforcement of intellectual property rights.

Service contracts are common. Mobile telephone service providers commonly require subscribers to sign up to a long term service contract in exchange for deeply-discounted mobile phones. The cost of the mobile phone is recovered from the subscriber over the length of the service contract. Some mobile phones even include an operator "lock-out" feature which prevents them from being used with a competitor's system.

Service expertise is also something with which nearly everyone is familiar. Automobile dealers can rely on service expertise to compensate for reduced margins on new car sales. All auto manufacturers provide special training and equipment to dealerships to help dealers generate aftermarket sales from service, maintenance, and repair. This option is however generally not available to low cost items. People are more tolerant of service when it comes to an automobile, a jet engine, or a copying machine, than they would be if the item was a cellular phone, or computer printer.

The *strategic use of technology* also acts to create, or cement, the dealer-customer service relationship. Many vehicle malfunctions can only be diagnosed, and repaired, by a licensed dealer with access to manufacturer-provided equipment. Similarly, the proprietary technology built into game consoles such as Microsoft's X-box™, or Nintendo's Gameboy™ locks the customer into purchasing game cartridges sold by the manufacturer.

Finally, aggressive *enforcement of intellectual property rights* (patents, trademarks, and copyrights) can be used to "lock-out" competitors. Then, as now, Gillette purposefully design and patent their razor and blade designs to ensure third-parties can't sell replacement blades without infringing their patents and then they aggressively enforce their rights through the courts. Owners of a Gillette razor are "locked in" to buying Gillette replacement blades because there are no compatible substitutes which can be manufactured or sold without infringing at least one (and preferably several) patent(s) owned by Gillette.

In short, Gillette's strategy was to lock-in customers and lock-out competitors. Customers were attracted by an innovative, high performance product - and lots of advertising. Durable manufacture and proper pricing of the company's safety razors created an installed base of satisfied customers. Enforcement of the company's patents ensured that replacement blades did not become a commodity.

From Razor Blades to Computer Printers

Gillette's model is a natural (and somewhat obvious) fit for the computer printer business where the state of the art technology is to have a replaceable ink-jet cartridge. Not surprisingly, it has been a successful strategy for companies like Hewlett Packard and Lexmark. According to *The Economist*, "*overpriced printer cartridges*" generate a substantial portion of  Hewlett-Packard's profits.[iii] Using essentially the same printer strategy as HP, Lexmark has grown from a small spin-off of IBM in 1991 to an international concern with $4.4 billion in sales by 2002.

In order to make the business a success, there are basically two tricks: the first is to convince the consumer that buying a printer which requires them to (frequently) pay €35 for a replacement printer cartridge is something they want to do, and the second is to prevent any competitor from selling cartridges, refilling kits, or remanufacturing services which would create a secondary market in replacement ink cartridges.

There has been a revolution in personal printing since Hewlett Packard introduced its first thermal ink-jet printer, called the ThinkJet, in 1984. Gone are the noisy dot matrix printers whose screaming print mechanisms provided the background music in many offices throughout the late 1970s and early 1980's. Thermal ink-jet printers are quieter and faster. Continual product improvement which increased ink dot densities and added color made thermal ink-jet printers became the printer of choice for a large number of printer users. The success which HP and Lexmark have enjoyed from their inkjet printer business is the result of a good business plan, inventive research and development, efficient commercialization of R&D results into useful consumer products, effective marketing, the continual development and release of new products, and a comprehensive patent strategy.

Patents, often maligned as legal rights which stifle innovation, have played – and continue to play - a central role to the success of the inkjet printer business. As I will describe in the following sections of this document, intellectual property rights are as important to the inkjet printer business as a steady supply of ink.

Whether or not patents stifle innovation is another subject. I will leave it the critics of the patent system to explain how it is NOT a progress of science that a photo quality computer printer can today be purchased for just over €100.

Chapter 2 - The Elements of a Patent Strategy

Before I discuss the patent strategies used by Hewlett Packard and Lexmark, it will be helpful to discuss patent strategy in general. For many executives patent strategy is the answer to the questions, "What inventions should we patent?" and "What patents should we license?" While the answers to these two questions may be adequate for some companies, most would benefit from a

Patent Strategy
^
Business

more comprehensive view of patent strategy. It is actually a mistake for me to use the phrase "patent strategy" because in my view there really is no such thing as a patent strategy - there is only a business strategy in which patents play a supporting (and sometimes decisive) role. In this regard, one could say that a comprehensive business strategy should consider the following patent related issues.

Patent (IPR) Issues to be considered in a Business Strategy

The following items are patent-related matters that often influence business activities. Not every item is applicable to every business strategy and there may be some items missing which are vital to other business strategies. Nevertheless, every business development effort should consider each of these items in detail.

1. The design of the product must be patentable.

Designing-in essential patented features which copiers must necessarily infringe is the *sine qua non* of Gillette's strategy. Designing a product which cannot be patented - or for which only weak patent protection is available - undermines the company's ability to protect its market position. If, as in the case of ink-jet printers, the business model hinges on the ability to prevent others from making, using, selling, new or remanufactured consumables it is vital that the consumables be protected by strong patents. Choices must be made during product development will directly influence the scope and strength of patent protection.

2. The design must (ideally) be free of the patents of others.

Designing-around existing patents to avoid unnecessary infringement is a vital aspect of any product design. Exposing a successful product to the (unexpected) payment of rent can devastate margins. A comprehensive patent clearance of new product designs is the best insurance against paying rent. While no product clearance is perfect, it is simply foolish to infringe patents which are easily identified and easily avoidable. Every business strategy must devote effort to identifying and designing around patents which cannot be licensed or for which only onerous license terms are available.

3. The design must consider trade secret protection

Trade secrets are the antitheses of patents. Patented inventions must be disclosed to the public, but it is not always technically or politically desirable for a company to make all of its technical features public knowledge. Failure to maintain a trade secret, such as revealing an encryption method, or making a product easy to hack, invites piracy. It is important to know what to patent and it is equally as important to know what should <u>not</u> be patented but kept as a trade secret.

4. The design must consider all other forms of IPR, including copyright and trademark.

Other IPR, trademark, copyright should always be considered in connection with the patent strategy. As I will discuss a bit later, Digital rights management (DRM) and the United States' Digital Millennium Copyright Act (DMCA) offer new avenues for product protection which must be considered.

5. There should be a plan for protecting the design by patent.

Patenting, the selection, filing, prosecution, and maintenance of the company's patent portfolio, is expensive. A single application for patent filed internationally can cost as much as €100,000 and a thicket of patents providing international protection for a product can cost more than €1 million. Patenting costs can, and usually do, consume a significant portion of R&D monies. The number, type, and geographical scope of patent protection must be balanced against the cost – and benefit – of the investment. There must also be a specific plan for prosecution so that patent claims are delivered which will be useful for enforcement and licensing.

6. There should be a plan (however preliminary) for how the company intends to enforce its patents.

Enforcement of the company's patents must begin before the patents are filed. Once a patent is issued, factors affecting its enforceability can be difficult, if not impossible, to correct. Patenting and product design must be done with an eye towards enforcement. Unless the company is able to successfully assert a patent, or group of patents, competitive advantage cannot be sustained. The stronger the patent position the less it will be necessary to rely on litigation. Patents which have claims which are clear, precise and which have been prosecuted in front of the most relevant prior art are powerful tools of negotiation. On the other hand, patents with broad or vague claims or which cite little, or no, prior art present significant difficulties for enforcement.

7. There should be a plan for licensing out the company's technology to others.

In some instances, the business plan may depend on making technology available to others. Sometimes, in order to build a market, it is necessary to make technology available to others. Making licenses available for patents used in open standards is absolutely necessary in order to achieve widespread adoption.

8. There should be a plan for licensing in technology needed from others.

In some instances, such as when third party patents cannot be designed around, the business plan may depend on acquiring licenses from others. Understanding precisely the scope of technology which is needed and negotiating license terms at and early stage guarantees margins will not be eroded. *Licensing-In* technology under favorable terms and conditions is vital to the success of the business strategy.

Failure to execute any of these elements can sink your business plan. Designing a product which cannot be patented - or for which only weak patent protection is available - undermines the company's ability to protect its market position. Failure to maintain a trade secret – or making products which are easy to hack into – invites piracy.

As you will see, in the case of ink-jet printers, Hewlett Packard and Lexmark have had to consider each of these elements.

Chapter 3 – Patent Strategies for Ink-Jet Printers

Before I begin this chapter, I must point out that this is not a treatise on thermal ink-jet printer technology, nor will I attempt here to present a comprehensive history of inkjet printer development. For the purpose of our study – which is to show how to use patents by illustrating some practical examples - it is sufficient to highlight aspects of inkjet printer technology and history.

In 1984 Hewlett Packard released its first thermal ink-jet printer for the consumer, called the ThinkJet™. *"The print cartridge used a glass substrate (instead of silicon), a nickel orifice plate and a rubber bladder for the ink delivery system."*[iv] Piezoelectric ink-jet printers, such as those manufactured by Epson, had a permanent print head built into the printer. In thermal ink-jet printers such as those manufactured by Hewlett Packard, Canon, Lexmark, Olivetti, Océ, and others, *"the print head is inside the ink cartridge, which includes all the microtechnology and an integrated circuit."*[v]

From the beginning, the commercialization of thermal ink-jet printer technology into a low cost cartridge – which was indeed HP's fundamental innovation - provided significant support for sustaining the overall ink-jet business model. The inherent complexity of thermal ink-jet cartridges meant that the possibility to design-in essential patents (and other intellectual property) would be far greater for thermal ink-jet printers than for piezoelectric technology where the ink cartridges supply only ink.

Immediately after the launch of the ThinkJet™, HP began filing the patents which would protect the ink-jet products which HP is selling today. In 1985, HP began shaping the patent/business strategy which contributes so much to their profits today and it is here that our analysis will begin.

It is interesting to note that some 10-15 years passed between when HP filed their initial patents and when they were used by HP to protect their market position. Even the ThinkJet™, released in 1984, was based on ink-jet technology developed by HP during the 1970's. Critics who argue that patent protection is "too long" should consider that many technologies take a decade or more after they are invented before they become commercially significant. It is not just pharmaceuticals which take a long time to come to market.

Development of the Patent Strategy

A patent strategy can only be developed in light of the overall business strategy so I will present this chapter by first developing an ideal "patent strategy" for ink-jet printers with respect to the issues discussed in the previous chapter. In the next chapter, I will attempt to illustrate aspects of how HP and, in certain cases, Lexmark adhere to this ideal strategy.

1. The design of the product must be patentable.

Designing-in is all about creating a patent position. Whether the patentable features which are designed into a product are the result of truly innovative advancements in technology, or minor ones, is of little consequence to the patent strategy. When it comes to ink jet printer cartridges, the product development strategy is a mixture of both true innovation and minor enhancements.

As good designs are the result of good specifications, the patent strategy begins with the product design specifications into which are added the following requirements:

a) The print cartridge design (and its physical and electrical interface to the printer) must itself be patentable.

b) The patentable features must be "essential" to the design. Essential means that one or more of the claims of a patent are necessarily infringed by a compatible replacement print cartridge.

c) The essential features of the print cartridge design are to be surrounded with a "thicket" of patents.

Due to the uncertainty attached to patent protection,

redundancy should be designed-in. A "thicket" of patents can mean a number of patents covering different aspects of a technological breakthrough, or better yet, a number of patents covering different design aspects of the ink jet printer cartridge.

In addition, the strategy for product design should consider ways to prevent re-manufacturing, or refilling of a used printer cartridge. Thus the design specifications may include the additional elements:

d) The ink cartridge should not be refillable, and/or the printer should not accept a refilled cartridge, and

e) All design documents relating to item d) shall be kept strictly confidential, or

f) The design of the ink cartridge should include patentable features essential to refilling, or remanufacturing, and

g) The essential refilling/remanufacturing features should be surrounded by another "thicket" of patents.

Another way of accomplishing this is to design the ink cartridge so that the printer head only works with a certain type of ink which the company can make proprietary.

The development task is thus not only to determine the best way for the company to produce the product and to prevent competition from compatible replacements, but to also consider how others may conceive to undermine the business model and to take preemptive steps to stop them.

2. The design must (ideally) be free of the patents of others.

Designing-around existing patents to avoid unnecessary infringement is a vital aspect of any product design, but it becomes increasingly difficult as a technology matures. Competitors seeking to create conditions for future cross licenses will build "picket fences" which may frustrate certain technology choices.

The risk of patent infringement is a usual and ordinary risk of business; the objective of designing-around is to reduce this risk as much as possible. As a first priority, patents owned by unwilling, or unreasonable licensors must be avoided, or dealt with at an early stage.

Thus, the design specifications should also include:

h) The printer and print cartridge design is to be checked for existing patents.

Since claim interpretation is foremost a legal matter, patent attorneys must assist in the product clearance. There is frankly some debate over whether or not it is advisable to conduct product clearance. "Why go looking for trouble?" appears to be an attitude which the law in some countries encourages. Nevertheless, it seems senseless to me to step on a mine when a cursory examination before one would prevent it. The depth and breadth of product clearance is something which each company must discuss with its legal advisers, and from a business perspective it is <u>always</u> something to be considered. If licenses are needed, they will always be easier to negotiate earlier in the development process, than later.

3. The design must consider trade secret protection

Trade secrets can provide protection against the ink cartridge becoming a commodity in countries where intellectual property rights are weak, or not respected. Trade secrets are however always vulnerable to reverse engineering. Consumer devices, because they are inexpensive and readily available, are particularly vulnerable to determined hacking. Trade secrets, such as embedding authentication codes in smart chips, may offer some short term advantage but will not replace strong patent protection. Trade secrets are generally not a panacea and may generate bad will among consumers, but they must always be considered.

4. The design must consider all other forms of IPR, including copyright and trademark.

In addition to the obvious notions of registering distinctive trademarks and designs, the technical complexity of ink-jet cartridges may offer the possibility to employ digital rights management (DRM) and the United States' Digital Millennium Copyright Act (DMCA). Whether or not such tools are used may be a matter of corporate policy, but these too should be considered. When changes to the law provide new avenues for intellectual property protection it would be foolish not to explore them. If you do not, your competitor will.

5. There should be a plan for protecting the design by patent.

There should be a target/budget for patenting. While some may argue that you cannot order inventions, this is poppycock. Inventions are the expected result of research and development and a portion of the R&D budget should be set aside for patenting these expected results.

The value of a patent may be difficult to determine, but the capital cost of acquiring patents is well documented in past invoices from law firms. The easiest way to create a budget for patenting is to begin by listing ideally what patents one would like to have, add up the numbers, and adjust upwards or downwards as necessary.

As with any other capital investment, the investment in patenting should be made proportional to expected future returns. A rule of thumb is that you don't have to protect every invention, but you do have to protect a sufficient number of inventions to protect your market position. One patent is generally risky, several patents on a particular feature is better, but best (and most costly) is to have several patents on each of several separate, distinct, and essential features.

Along with the first prototype sketches, there should be a plan for the number and type of patents which will be sought. Ideally, patents should be sought in the following areas:

- technical features in the printer,
- technical features in the ink cartridge, and
- technical features in the interface between the printer and the ink cartridge.

The technical features which are patented should be:

- essential to the operation or manufacture of the printer, or print cartridge,
- visible to the user, or obvious with minimal experimentation,
- technically simple.

Inventions received from R&D can be "judged" along these parameters and patents filed in accordance with the allocated budget.

Selecting the right inventions is only part of the process. There must also be a specific plan for prosecution of the applications which are filed so that patent claims are delivered which will be useful for enforcement and licensing. Selection and prosecution cannot be black boxes – the patent attorney must understand why an invention was selected and work towards that goal.

For example, seeking the broadest possible patent coverage is undesirable if the goal is simply to prevent compatible inkjet cartridges from appearing on the market. Almost without exception infringers make two defenses: 1) they are not infringing the claims, and 2) the claims are invalid. Infringement is easiest to prove when the claims are essential, visible to the user, and technically simple. Invalidity is hardest to prove when the claims are narrow, detailed, and specific. In general one can say the broader the claim, the greater the scope of prior art which could be found to invalidate it. This is not to say that broad claims should not be sought during prosecution, but merely to emphasize that the prosecuting attorney's first obligation is to obtain claims to support the business model.

6. There should be a plan (however preliminary) for how the company intends to enforce its patents.

Enforcement of the company's patents must begin before the patents are filed. Once a patent is issued, factors affecting its enforceability can be difficult, if not impossible, to correct. Patenting and product design must be done with an eye towards enforcement. The stronger the patent position the less it will be necessary to rely on litigation. Patents which have claims which are clear, precise and which have been prosecuted in front of the most relevant prior art are powerful tools of negotiation. On the other hand, patents with broad or vague claims or which cite little, or no, prior art present significant difficulties for enforcement.

In the case of ink-jet cartridges, what patents (or, more precisely, patent claims) would the company need to enforce against manufacturers, re-manufacturers, or sellers of refill kits.

7. There should be a plan for licensing out the company's technology to others.

In order to ensure compatibility with different operating systems and platforms, driver software must be licensed and made available on suitable terms.

8. There should be a plan for licensing in technology needed from others.

In some instances, such as when third party patents cannot be designed around, the business plan may depend on acquiring licenses from others. Understanding precisely the scope of technology which is needed and negotiating license terms at and early stage guarantees margins will not be eroded. *Licensing-In* technology under favorable terms and conditions is vital to the success of the business strategy.

Chapter 4 – Examples

"Hewlett-Packard Company (NYSE:HWP) today announced that it has won a significant patent lawsuit relating to thermal inkjet printer cartridges.

... HP's complaint contended that the companies infringed six HP patents by importing or selling inkjet cartridges manufactured by Microjet, including one model intended to replace the HP 51626A cartridge and another model intended to replace the HP 51629A cartridge.

The ITC administrative law judge issued a lengthy decision detailing how the Microjet cartridges infringe five of the six HP patents. The judge also resoundingly rejected Microjet's claims that the patents were invalid." [vi]

- HP Press Release, Jan. 28, 2002

Overpriced Printer Cartridges

Enforcement is where the rubber meets the road. What patents are enforced, how they are enforced, and against whom, illustrates a great deal about the patent strategy. In this chapter, I will attempt to present examples of how HP and, as the case may be, Lexmark have executed the patent strategy which I discussed in the previous chapter. Let's begin with examining some aspects of the design.

1. The design of the product must be patentable.

Table 2 illustrates the patents which HP uses to protect some of their more popular ink jet printer cartridges.[vii] HP 51625A and HP 51649A are older designs compared to HP 51626A and HP 51629A.

As I mentioned previously, whether the patentable features which are designed into a product are the result of truly innovative advancements in technology, or minor ones, is of little consequence to the patent strategy. Product development is often a mixture of both true innovation and minor enhancements.

US Patent Number	HP Part Number			
	HP51625A	HP51626A	HP51629A	HP51649A
4,635,073	x	x	x	x
4,680,859	x	x	x	x
4,683,481	x			x
4,771,295	x			x
4,827,294		x	x	
4,872,027		x	x	
4,992,802		x	x	
5,409,134		x	x	

Table 2 – U.S. Patents used in certain HP Print Cartridges

A listing of the titles illustrates the different aspects of the print cartridge design and manufacture which have been covered by U.S. Patent - and also shows the progression of design choices which have been made. The first two patents listed in Table 1 are common to all four of the print cartridges shown:

> U.S. Patent 4,635,073 Replaceable thermal ink jet component and thermosonic beam bonding process for fabricating same

> U.S. Patent 4,680,859 Thermal ink jet print head method of manufacture

These two patents cover fundamental technology which is built-into the ink-jet printer cartridge. Although one of these patents would, in theory, be sufficient to prevent competitors from making, using, or selling compatible print cartridges, redundancy, in the form of patents on improvements to the fundamental technology were also patented.

> U.S. Patent 4,683,481 Thermal ink jet common-slotted ink feed printhead

> U.S. Patent 4,771,295 Thermal ink jet pen body construction having improved ink storage and feed capability

By covering various aspects of the design and manufacture, Hewlett Packard created a "thicket" of patents on their early print cartridges.

One unavoidable problem with relying on patents is that most, if not all, patents expire 20 years after the earliest date of priority. Since the fundamental patents (U.S. Patent 4,635,073 and U.S. Patent 4,680,859) are due to expire in 2005 and 2006 respectively, newer cartridges were designed with additional patented features to effectively extend their patent protection beyond the expiration date of the fundamental patents.

Extending Patent Protection

Three of these patents reflect tangible improvements in ink jet printer technology:

> U.S. Patent 4,827,294 Thermal ink jet printhead assembly employing beam lead interconnect circuit

> U.S. Patent 4,992,802 Method and apparatus for extending the environmental operating range of an ink jet print cartridge

> U.S. Patent 5,409,134 Pressure-sensitive accumulator for ink-jet pens

One patent, however, is highly illustrative of the steps HP used to extend their monopoly on replacement print cartridges.

Inventing an Invention

U.S. Patent 4,872,027 (abbreviated hereinafter as the Buskirk '027 patent) is an excellent example of how Hewlett Packard has engineered patents to protect their competitive advantage. The abstract of the Buskirk '027 patent describes the invention as:

> *"A dot-matrix printer is provided with different types of printheads which are interchangeably attachable to the printer carriage. The heads are provided with individual codes which are read by the printer control system and used to reconfigure its control function to suit the control requirements of the identified head. Such a system may include a microprocessor responsive to individual sets of instructions or programs providing new and different processing capabilities for printing control in response to the insertion of a new printhead."[viii]*

The importance of this invention to Hewlett Packard's business strategy is revealed in the complaint filed with the United States

International Trade Commission by Hewlett Packard on December 22, 2002.[ix]

> *"32. In general, the Buskirk '027 patent is directed to an innovative identification scheme for ink jet cartridges. Before the Buskirk invention, ink jet printers were not able to automatically identify which type of print cartridge a user had installed in the printer. Different cartridges may have identical external characteristics (such as size, shape, and physical contour) but may nevertheless have very different printing parameters (such as different colored inks, firing frequencies, dot sizes, etc.). **The printer must be able to identify the precise parameters of a particular cartridge, so that the printer can operate correctly for that particular cartridge (or give a signal to the user that an improper one has been installed)**"*[x] (emphasis added.)

From this it is may be assumed that the main printer assembly is equipped with circuitry to identify the print cartridge – which of course is a very convenient way of ensuring that the printers only work with authorized print cartridges. Since the printer expects to see a certain resistance (and will not operate without this) compatible printer cartridges must necessarily infringe the patent. This is what is meant by an essential feature.

A Trivial Invention

An examination of claim 4 of the Buskirk patent reveals that this is an absolutely trivial invention.

> *4. Thermal inkjet printhead identification means, comprising;*
>
> *a printhead body having at least one ink chamber;*
>
> *a nozzle plate on said printhead body having nozzles communicating with said chamber;*
>
> *a resistor network having an ink expulsion resistor at each nozzle;*
>
> *contact pads in said resistor network and individual circuits connecting individual contact pads to individual resistors; and*
>
> *at least two printhead identification contact pads, each disposed*

between selected different pairs of said contact pads, and forming part of a printhead identification resistor network including at least two resistors for each printhead identification contact pad.

The "invention" is the addition to an existing ink jet printer cartridge of a pair of contact pads presenting different resistance values to a "head identity" circuit embedded in the printer's microprocessor: an embarrassingly simple solution which nevertheless (apparently) satisfies all of the statutory requirements for a patent.

An Open and Shut Case for Infringement

Without these contacts, "compatible" ink cartridges would not be compatible, and with them, Hewlett Packard's patent is necessarily infringed. More importantly, this is a feature which is easily described, visibly obvious, and simply verified by the use of an ohmmeter. Few courts would have difficulty understanding this patent, or the evidence demonstrating its infringement.

2. The design must (ideally) be free of the patents of others.

It is impossible for an outsider (such as myself) to know what, if any, measures HP employed to avoid third party patent infringement, except that it would appear that it was not possible for HP to avoid the patents of at least one of their competitors...

> *"Lexmark International, Inc. today announced it has signed an agreement with Hewlett-Packard Company to cross-license each other's patents filed prior to a specified date.*
>
> *While the specific details of the agreement are confidential, the agreement generally gives the companies a worldwide license under the licensed patents for the manufacture and sale of printers, as well as accessories and consumable supplies designed for use with each company's own printers. The agreement resolves issues of patent infringement that had been raised by both companies and does not involve any royalty or other payments by either party."*

- Oct 21 1996 Press Release from Lexmark International, Inc.

3. The design must consider trade secret protection

There are trade secrets which are technically necessary (such as not revealing encryption codes) and trade secrets which are politically necessary to protect the reputation of the company.

Purposefully designing an ink cartridge so that it is not refillable, or a printer so that it will not accept a refilled cartridge may not be the sort of positive product message a company wants to generate. As part of the patent strategy, sensitive areas such as this which could affect the company's reputation should be discussed with responsible people from the communications department and a policy for confidentiality created.

One way of preventing a secondary market from emerging would be to insert "secret" one-time-use features into a design (such as codes embedded in smart cards) which cannot easily be discovered or defeated by unlicensed replacements. While such a function may also be politically sensitive, trade secret protection is nevertheless necessary to preserve its technical function.

Part of Lexmark's trade secret strategy was recently revealed when Lexmark filed a complaint with the US Federal Court in Lexington, Kentucky alleging that a competitor, Static Control Components, was violating the Digital Millennium Copyright Act (DMCA) by selling a microchip which enables non-Lexmark replacement cartridges to work with Lexmark printers.

According to the complaint, Lexmark printers and toner cartridges use a cryptographic "secret handshake" to authenticate each other. Lexmark printers refuse to accept a cartridge which does not know the handshake.[xi]

The trade secret, as is often the case, was subjected to reverse engineered and wasn't a secret for very long. Nevertheless, every patent strategy should consider that having trade secret protection on top of patent protection further strengthens the company's proprietary position. As you will see, Lexmark's successful use of the US DMCA added yet another dimension to their protection.

4. The design must consider all other forms of IPR, including copyright and trademark.

The "patent" strategy should also consider other forms of intellectual property protection. Copyright of proprietary drivers and use of proprietary fonts are common restrictions used in the print industry.

Creative use of copyright law was also revealed by Lexmark's lawsuit. According to *The Register*,

> *"Lexmark's complaint alleges that Smartek microchips incorporate infringing copies of its software and are being sold by Static Control to defeat Lexmark's technological controls, hence invocation of the DMCA. Its case is that Static Control's technology permits the unauthorized remanufacturing of Lexmark Prebate toner cartridges."[xii]*

Despite howls of outrage from opponents that the DMCA was never intended for such use, the court agreed with Lexmark.

> *"As a result of a Lexmark International, Inc. (NYSE: LXK) lawsuit against Static Control Components, Inc., for violation of the Copyright Act and the Digital Millennium Copyright Act, the federal district court in Lexington, Ky., issued a temporary order – agreed to by Static Control – requiring Static Control to immediately cease making, selling, or otherwise trafficking in the "Smartek™" microchip for the toner cartridges developed for the Lexmark T520/522 and T620/622 laser printers."[xiii]*

- Jan. 9th, 2003 Press Release from Lexmark International, Inc.

This pioneering use of the DMCA shows that Lexmark is on the cutting edge when it comes to making the most of all aspects of intellectual property law. Notably, among the critics of Lexmark for using the DMCA was HP:

> *"A top Hewlett-Packard printer executive said that although intellectual property rights are vital in the printer industry, rival Lexmark is wrong to try to use a controversial copyright law to safeguard those rights.*
>
> *In an interview with CNET News.com, HP Senior Vice President Pradeep Jotwani said Lexmark is using the 1998 Digital Millennium Copyright Act in ways it was*

not intended, in pursuing a lawsuit against a maker of remanufactured toner cartridges.

"We think it is stretching it," Jotwani said of the Lexmark suit, which was filed Dec. 30 against Static Control Components, a maker of remanufactured toner cartridges. "The DMCA was put in place (to protect) things like movies, music and software applications."

Jotwani said HP will protect its intellectual property rights if companies infringe on them, but the DMCA is not the right weapon to use. A Lexmark representative declined to comment specifically on Jotwani's remarks.

"I don't plan on going down that path," Jotwani said, referring to using the DMCA."[xiv]

- CNET News.com Feb. 5th, 2003

After Lexmark's victory in the Federal Circuit, "*that path*" just got wider and more well-defined.

Score: Lexmark 1, HP 0

In the final analysis, the opinion of company managers about the DMCA is far less important than the opinion of the US Federal Courts. Despite the derision from HP, the US Federal Court did not think that Lexmark's use of the DMCA was "*stretching it*" and agreed that the US DMCA can be used to protect ink-jet cartridges precisely as Lexmark used it. When one considers that copyright protection extends more than 3 times as long as patent protection, and that the anti-circumvention clause has successfully been employed by a competitor, HP's "patent" rejection of the DMCA as a tool to protect its printer business seems reckless.

5. There should be a plan for protecting the design by patent.

HP has been issued approximately 174 US Patents related to thermal ink-jet printers resulting from applications filed with the US Patent and Trademark Office (USPTO) between November 22nd, 1985 and December 11th, 1991.[xv] Why do I choose these dates? Well, because this is the time span between the oldest and newest of the 6 US patents were asserted against Microjet in the aforementioned ITC action.

What is interesting to note is that out of (approximately) 174 patents related to thermal ink-jet printers, 6 were decisive. This is not to say that the other 168 patents are (were) unnecessary, but it does suggest that HP's strategy for patenting during this time was to file a lot of applications for patent related to ink-jet printers. On one hand, one can look at it like this: if 174 patents were needed to produce 6 winners, and those 6 winners defend a multi-billion dollar business, then it was a good investment. On the other hand - while there is definitely some uncertainty with patents and patenting – very few companies can afford the inefficiency of such a strategy.

There should be a target/budget for patenting. Spend as much as you can afford, or as little as you dare, but have a budget - and a plan for how the budget will be spent.

A plan for prosecution

The goal of patent drafting and prosecution is to obtain claims which support the business model. In other words, the goal of prosecution is identical to the goal of product design. Patent prosecution should seek to obtain claims which are:

    i. Essential to the product
   ii. Easily described
  iii. Visibly obvious
  iv. Simply verified

Plus one additional, but highly important, aspect:

    v. Sufficiently detailed and narrow as to be difficult to invalidate.

5 out of 6 of HPs patents were found to be infringed by the ITC administrative law judge; the judge also "*resoundingly rejected*" Microjet's claims that the patents were invalid.

A litigator will always have a better chance with a single independent claim, clearly written than with a dependent claim written using obscure and complicated terms. And when your litigator has a better argument, <u>you</u> have a better argument.

6. There should be a plan (however preliminary) for how the company intends to enforce its patents.

It is usually not difficult to imagine what competitors will do to grasp a share of a successful market. In the case of ink-jet printers, competition is most likely to come from manufacturers of "compatible" replacement ink cartridges, industrial re-manufacturing of inkjet cartridges, or re-filling kits sold to individuals.

The company must decide even before patents are filed against whom the patents would be enforced, and what sort of patents would be most effective. It is not enough to simply obtain patents on a product and hope that these will be useful against competitors. One must have - in advance - an idea of how these patents will be used and against whom.

While it may be advantageous to allow ink-jet cartridges to be refilled a number of times by individual users, large scale industrial re-manufacturing would likely undermine the entire business model. Patents, or more precisely patent claims, which could be used in an enforcement action against re-manufacturers, or their retailers, are vital to success.

7. There should be a plan for licensing out the company's technology to others.

Both Lexmark and HP make free licenses available for driver software for MacOS, Microsoft Windows, and Linux.

8. There should be a plan for licensing in technology needed from others.

Unlike product clearance which tries to avoid unnecessary infringement of third party patents, some third party patents must necessarily be infringed. Products, like computer printers, need to be compatible with a wide range of computers or network servers. Towards this end, open standards such as IEEE 1394 have been developed. Lexmark is a licensee of the 1394LA, the licensing authority established to provide fair, reasonable, nondiscriminatory worldwide access under a single license to patents that are essential for implementing the international standards relating to a high speed transfer digital interface known as IEEE 1394 (which Apple calls "Firewire" and Sony calls "i.Link.")

A careful mapping of what licenses are needed – and how much they will cost – is an essential part of the patent strategy.

Execution of the Strategy: Patent Engineering

Rather than patenting inventions which spontaneously arise from development work (the old-fashioned way in which most patents are generated) the patent strategy provides guidance to the development team to engineer patents which will support the business strategy.

In other words, the development task is more than simply developing a working design and then filing the patents to protect it (the normal way R&D has historically been accomplished.) By specifying exactly what sort of patents which are needed, the development effort is focused on engineering a product which satisfies critical patent requirements. In complex technologies, the patent strategy requires that patents become an integral part of the entire business strategy and not just a stand-alone subsection.

Chapter 5 – Adapting to Competition

"But the threat to HP's ink business is increasing. Recently, makers such as Singapore-based INKE have begun selling desktop automatic ink refill stations costing US$49. The company claims savings of US$329 per average user over the three-year lifespan of an inkjet printer. INKE ink tanks cost just US$6 each, compared with HP's original cartridges, which can cost up to US$30."[xvi]

- CNET Asia, October 20th, 2003

With some observers noting that, ounce for ounce, inkjet printer ink costs more than 1985 Dom Perignon[xvii], it should come as no surprise that the business invites fierce competition. Just recently, Inké (www.inke.com.sg) proudly unveiled its inaugural product, the Inké HS-45 inkjet auto refill system, at Comex 2003 (held from 4-7 September 2003 at the Singapore International Convention and Exhibition Centre at Suntec City.)

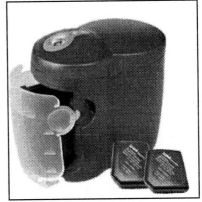

Automated re-filling machines such as Inké's HS 45 present a competitive challenge to HP's business model. Widespread use, especially among small businesses, could cut severely into HP's aftermarket sales of ink-jet printer cartridges. Due to patent exhaustion, refilling ink-jet cartridges does not generally require any additional license to HP's formidable patent portfolio. Unable to threaten patent infringement, HP might be forced to cut prices....

Staying one step ahead

Well-run companies do not sit on their hands waiting for the competition to move. HP is a well-run company. More than two years before Inké's unveiling, HP had already filed a series of patents to thwart this challenge. Conspicuously avoiding the use of the word "refill," HP filed a series of patents on its own "refilling" unit which HP attorneys termed a "rejuvenation station."

> *When asked why HP would try to patent a device that could hurt its own disposable cartridge business, Christina Tay, a vice-president at ink-refill device maker INKE said that "it is difficult to speculate on their motive".[xviii]*

It is difficult to speculate about HP's motive? Well, it shouldn't be. HP's motives are probably the same as those of any other market leader: to maintain its market share and high margins – and to drive its competitors out of business.

Planned obsolescence

A perusal of the claims of the granted patent, and pending applications listed above, indicates why HP is a well-run company.

Patent Number	Title	Publication Application Number	Filed	Related US Patent Application Data
6,478,415	Rejuvenation station and printer cartridge therefore	2002/0135645	March 21, 2001	
Pending	Rejuvenation station and printer cartridge therefore	2003/0011664	September 10, 2002	Division of 09/814,329 now US Patent 6,478,415
Pending	Rejuvenation station and printer cartridge therefore	2003/0011665	September 9, 2002	Division of 09/814,329 now US Patent 6,478,415
Pending	Rejuvenation station and printer cartridge therefore	2003/0011666	September 9, 2002	Continuation of Application 09/814,329 now US Patent 6,478,415

Table 5.1 – Recently granted HP patents and pending applications on a "Rejuvenation station and printer cartridge therefore."

The granted US patent and pending US applications describe an ink-jet cartridge having two refillable ink reservoirs with a careful balance of pressure between them. In addition to (allegedly) improving the print quality, this arrangement also makes the ink-jet cartridge difficult to refill using a syringe, or automatic filling station. Moreover, refilling will be difficult, if not impossible, without infringing one of more claims pending before the United States Patent and Trademark Office (USPTO) which include claims covering:

- "a rejuvenator station"
- "a printer cartridge"
- "a process of refilling"
- "printer cartridge and a rejuvenation station"
- "a method of pumping fluid"
- "a method to stop pumping fluid"

The patent and pending applications suggest that future ink jet cartridges will undergo a design change from having a single reservoir which is easily refilled to a construction having two refillable ink reservoirs. In other words, by the time Inké is able to cut into HP's business, the business will already be somewhere else and Inké (and others) will probably not be able to construct new refilling stations without infringing HP's new patents. Unable to work with new HP catrtidges, Inké's products will made obsolete.

Cautions

This document is based entirely on publicly available information, referenced and footnoted herein. I have never worked for, nor have any "inside information" about either Hewlett Packard or Lexmark. As such, I have necessarily made some assumptions which may, or may not, be accurate. The purpose of this document is solely to illustrate the practical use of patents as part of a business strategy and the examples presented herein are exemplary only.

Patent Strategy, because it deals primarily with the exercise of monopoly rights, must always be considered with due regard to applicable intellectual property and anti-competition laws. These laws vary substantially between nations and regions. The reader of this document is cautioned to obtain appropriate legal advice before executing any patent strategy based on the contents of this document.

References

i Mansfield, J., The Razor King. *American Heritage of Invention and Technology*, Spring 1992, pp. 40-46.

ii Gillette never faced the problem of flat sales. In 1904, Gillette sold 51 safety razors, in 1905, nearly 39,000, and by 1908, annual sales of Gillette safety razors topped 1 million.

iii Merger Mystery, *The Economist*, November 16th, 2002

iv http://www.hp.com/oeminkjet/learn/beginnings.html

v 2002, *Superhot Dots*, Scientific American, December Issue

vi http://www.hp.com/hpinfo/newsroom/press/28jan02b.htm

vii Patent numbers are printed on Hewlett Packard's packaging of for these model number print cartridges.

viii Abstract of U.S. Patent 4,872,027

ix Documents filed in support of Hewlett-Packard's request that the Commission commence and investigation pursuant to section 337 of the Tariff Act of 1930, as amended 19 U.S.C. § 1337, December 22, 2000. Available on-line at http://www.usitc.gov

x Ibid, page 12.

xi This strategy does not work everywhere. In an effort to reduce European landfill waste, the European Union has recently banned the use of "clever chips" which ensure that ink cartridges cannot be refilled.

xii "Lexmark unleashes DMCA on toner cartridge rival," *The Register*, January 10th, 2003.

xiii"Lexmark lawsuit seeks to defend intellectual property rights while preserving customers' rights to choose," Lexmark Press Release, January 9th, 2003
http://www.lexmark.com/US/press_releases_details/0,1233,Nzg2fDE=,00.html

xiv "HP raps rival for invoking DMCA," *CNET News.com*, February 5, 2003
http://news.com.com/2100-1040-983518.html

xv A search performed on Sept. 19th, 2003 of USPTO (http://www.uspto.gov) records using the Boolean string: AN/Hewlett-Packard and APD/11/22/1985->12/11/1991 and ABST/((ink and jet) or inkjet) returned 174 issued US patents.

xvi "HP files patents for ink-refill device" October 20, 2003 http://asia.cnet.com/newstech/personaltech/0,39001147,39155063,00.htm

xvii http://www.gizmodo.com/archives/008611.php

xviii "HP files patents for ink-refill device" October 20, 2003 http://asia.cnet.com/newstech/personaltech/0,39001147,39155063,00.htm

Printed in the United States
37111LVS00003B/629-640